YOUR KNOWLEDGE HAS VALUE

AF388618

- We will publish your bachelor's and
  master's thesis, essays and papers

- Your own eBook and book -
  sold worldwide in all relevant shops

- Earn money with each sale

Upload your text at www.GRIN.com
and publish for free

Bibliographic information published by the German National Library:

The German National Library lists this publication in the National Bibliography;
detailed bibliographic data are available on the Internet at http://dnb.dnb.de .

Imprint:

Copyright © 2016 GRIN Verlag, Open Publishing GmbH
Print and binding: Books on Demand GmbH, Norderstedt Germany
ISBN: 9783668351240

This book at GRIN:

http://www.grin.com/en/e-book/344950/the-role-of-phytoremediation-in-remedia-
tion-of-industrial-waste

Yunusa Adamu Ugya, T. S. Imam, S. M. Tahir

The Role of Phytoremediation in Remediation of Industrial Waste

GRIN Publishing

The Role of Phytoremediation in Remediation of Industrial Waste

Abstract

This chapter summarizes the role of phytoremediation in the remediation of industrial waste water since this waste water has become a treat to water quality. Several technologies are available to remediate water that is contaminated by industrial pollutant. However, many of these technologies are costly (e.g. excavation of contaminated material and chemical/physical treatment) or do not achieve a long-term nor aesthetic solution. Phytoremediation can provide a cost-effective, long-lasting and aesthetic solution for remediation of contaminated sites. In many cases, especially in tropical or subtropical areas, invasive plants such as the water hyacinth (*Eichhornia crassipes*) and water lettuce (*P. stratiotes* L.) are used in these phytoremediation water systems. This is because, compared to native plants, these invasive plants show a much higher nutrient removal efficiency with their high nutrient uptake capacity, fast growth rate, and big biomass production. In the active growth season, for instance, water hyacinth plants can double in number and biomass in 6 to 15 days. This study shows the important of phytoremediation in the phytoremediation of industrial waste.

Keywords: *Lemna minor, Pistia stratiotes, Eicchornia crassipes, Azolla pinnata,* Waste water

Content

1 Introduction

The principles of phytoremediation system are to clean up contaminated water, which include identification and implementation of efficient aquatic plant; uptake of dissolved nutrients and metals by the growing plants; and harvest and beneficial use of the plant biomass produced from the remediation system (Lu, 2009). The most important factor in implementing phytoremediation is the selection of an appropriate plant (Roonngtanakiat *et al.*, 2007; Stefani *et al.*, 2011), which should have high uptake of both organic and inorganic pollutants, grow well in polluted water and easily controlled in quantitatively propagated dispersion (Roonngtanakiat *et al.*, 2007). The uptake and accumulation of pollutants vary from plant to plant and also from specie to specie within a genus (Singh *et al.*, 2003). The economic success of phytoremediation largely depends on photosynthetic activity and growth rate of plants (Xia and Ma, 2006), and with low to moderate amount of pollution (Jamuna and Noorjahan, 2009). Many researchers have used different plant species like water hyacinth *(Eichhornia crassipes (Mart.) Solms)* (Muramoto and Oki, 1983; Trivedy and Pattanshetty, 2002; Mahmood *et al.*, 2005; Dhote and Dixit, 2007; Jamuna and Noorjahan, 2009; Lissy *et al.*, 2010; Valipour *et al.*, 2010; Valipour *et al.*, 2011; Dar *et al.*, 2011;), Water Lettuce (*Pistia stratiotes* L.)(Fonkou *et al.*, 2002; Jing *et al.*, 2002; Awuah *et al.*, 2004; Lu *et al.*, 2010; Dipu *et al.*, 2011), Duckweed *(Water Lemna)*, Bulrush *(Typha)*, Vetiver Grass *(Chrysopogon zizanioides)* (Troung and Baker, 1998; Lakshmana *et al.*, 2008; Girija *et al.*, 2011) and Common Reed *(Phragmites Australis)* for the treatment of water. They have used these species for different types of contaminated waters, effluents etc. Mkandawire and Dudel (2007) have used duckweed and they found its growth was restricted above 34 °C and pH sensitive. Mashauri *et al.*, (2000) used bulrush and his study revealed that the Total Dissolved Solids (TDS) and Electrical Conductivity (EC) concentration was increased after treatment. Baskar *et al.* (2009) in his study of kitchen wastewater treatment found only 4% TDS removal by common reed.

The waste water released from industries are characterized by presence of large quantity of polycyclic and aromatic hydrocarbon, phenols, metal derivatives, surface active substances, sulfides, naphthalene acids and other chemicals (Suleimanov, 1995). As a result of ineffectiveness in the purification systems, this waste water lead to the accumulation of toxic products in the receiving waste water bodies with potentially consequences on the ecosystem (Aghalino and Eyinla, 2009).

2 Water Quality: A Worldwide Concern

"Water quality" is a term used to express the suitability of water to sustain various uses or processes. Any particular use will have certain requirements for the physical, chemical or biological characteristics of water; for example limits on the concentrations of toxic substances for drinking water use, or restrictions on temperature and pH ranges for water supporting invertebrate communities. Consequently, water quality can be defined by a range of variables which limit water use. Although many uses have some common requirements for certain variables, each use will have its own demands and influences on water quality (McJunkin, 1982; Hemp, 1984; Raskin and Ensley, 2000).

Quantity and quality demands of different users will not always be compatible, and the activities of one user may restrict the activities of another, either by demanding water of a quality outside the range required by the other user or by lowering quality during use of the water. Efforts to improve or maintain a certain water quality often compromise between the quality and quantity demands of different users. There is increasing recognition that natural ecosystems have a legitimate place in the consideration of options for water quality management. This is both for their intrinsic value and because they are sensitive indicators of changes or deterioration in overall water quality, providing a useful addition to physical, chemical and other information (Serruya and Pollingher, 1983; WHO, 1993).

The composition of surface and underground waters is dependent on natural factors (geological, topographical, meteorological, hydrological and biological) in the drainage basin and varies with seasonal differences in runoff volumes, weather conditions and water levels (Foster and Hirata, 1988).

Large natural variations in water quality may, therefore, be observed even where only a single watercourse is involved. Human intervention also has significant effects on water quality. Some of these effects are the result of hydrological changes, such as the building of dams, draining of wetlands and diversion of flow. More obvious are the polluting activities, such as the discharge of domestic, industrial, urban and other wastewaters into the water course (whether intentional or accidental) and the spreading of chemicals on agricultural land in the drainage basin (Serruya and Pollingher, 1983).

Water quality is affected by a wide range of natural and human influences. The most important of the natural influences are geological, hydrological and climatic, since these affect the quantity and the quality of water available. Their influence is generally greatest when available water quantities are low and maximum use must be made of the limited resource; for example, high salinity is a frequent problem in arid and coastal areas (Hemp, 1984).

If the financial and technical resources are available, seawater or saline groundwater can be desalinated but in many circumstances this is not feasible. Thus, although water may be available in adequate quantities, its unsuitable quality limits the uses that can be made of it.

Although the natural ecosystem is in harmony with natural water quality, any significant changes to water quality will usually be disruptive to the ecosystem.

The effects of human activities on water quality are both widespread and varied in the degree to which they disrupt the ecosystem and/or restrict water use. Pollution of water by human faeces, for example, is attributable to only one source, but the reasons for this type of pollution, its impacts on water quality and the necessary remedial or preventive measures are varied. Faecal pollution may occur because there are no community facilities for waste disposal, because collection and treatment facilities are inadequate or improperly operated, or because on-site sanitation facilities (such as latrines) drain directly into aquifers (WHO, 1988).

The effects of faecal pollution vary. In developing countries intestinal disease is the main problem, while organic load and eutrophication may be of greater concern in developed countries (in the rivers into which the sewage or effluent is discharged and in the sea into which the rivers flow or sewage sludge is dumped). A single influence may, therefore, give rise to a number of water quality problems, just as a problem may have a number of contributing influences. Eutrophication results not only from point sources, such as wastewater discharges with high nutrient loads (principally nitrogen and phosphorus), but also from diffuse sources such as run-off from livestock feedlots or agricultural land fertilised with organic and inorganic fertilisers. Pollution from diffuse sources, such as agricultural runoff, or from numerous small inputs over a wide area, such as faecal pollution from settlements, is particularly difficult to control (McJunkin, 1982).

The quality of water may be described in terms of the concentration and state (dissolved or particulate) of some or all of the organic and inorganic material present in the water, together with certain physical characteristics of the water. It is determined by *in situ* measurements and by examination of water samples on site or in the laboratory. The main elements of water quality monitoring are, therefore, on-site measurements, the collection and analysis of water samples, the study and evaluation of the analytical results, and the reporting of the findings. The results of analyses performed on a single water sample are only valid for the particular location and time at which that sample was taken. One purpose of a monitoring programme is, therefore, to gather sufficient data (by means of regular or intensive sampling and analysis) to assess spatial and/or temporal variations in water quality (Meybeck *et al.*, 1989; Nash and McCall, 1994). The quality of the aquatic environment is a broader issue which can be described in terms ofwater quality,the composition and state of the biological life present in the water body,the nature of the particulate matter present, andthe physical description of the water body (hydrology, dimensions, nature of lake bottom or river bed, e.t.c

Complete assessment of the quality of the aquatic environment, therefore, requires that water quality, biological life, particulate matter and the physical characteristics of the water body be investigated and evaluated. This can be achieved throughchemical analyses of water, particulate matter and aquatic organisms (such as planktonic algae and selected parts of organisms such as fish muscle),biological tests, such as toxicity tests and measurements of enzyme activities, descriptions of aquatic organisms, including their occurrence, density, biomass, physiology and diversity (from which, for example, a biotic index may be developed or microbiological characteristics determined), and physical measurements of water temperature, pH, conductivity, light penetration, particle size of suspended and deposited material, dimensions of the water body, flow velocity, hydrological balance, etc (Foster and Gomes, 1989; Meybeck and Helmer, 1996).

Pollution of the aquatic environment, as defined by GESAMP (1988), occurs when humans introduce, either by direct discharge to water or indirectly (for example through atmospheric pollution or water management practices), substances or energy that result in deleterious effects such ashazards to human health,harm to living resources,hindrance to aquatic activities such as fishing,impairment of water quality with respect to its use in agriculture, industry or other economic activities, or reduction of amenity value. The importance attached to quality will depend on the actual and planned use or uses of the water (e.g. water that is to be used for drinking should not contain any chemicals or microorganisms that could be hazardous to health). Since there is a wide range of natural water qualities, there is no universal standard against which a set of analyses can be compared. If the natural, pre-polluted quality of a water body is unknown, it may be possible to establish some reference values by surveys and monitoring of unpolluted water in which natural conditions are similar to those of the water body being studied (Nash and McCall, 1994).

2.1 Impacts of Water pollution

The extent of anthropogenic environmental pollution in the developing world is well documented (Mattina *et al.*, 2003). Among overall environmental pollution, water pollution is one of the major threat to public health especially in developing and under developed countries as drinking water quality in these countries is poorly managed and monitored (Mwegoha, 2008; Azizullah *et al.*, 2011). Both surface and ground drinking water get contaminated with coli forms, toxic metals and pesticides. About 2.3 billion peoples are suffering from water related diseases worldwide (UNESCO, 2003). The presence of heavy metals (elements with an atomic density greater than 6 g/cm) is one of the most persistent pollutants present in water. Unlike other pollutants, they are difficult to degrade, but can accumulate throughout the food chain, producing potential human health risks and ecological disturbances (Akpor and Muchie, 2010). In developing countries, more than 2.2 million people die every year due to drinking of contaminated water and inadequate sanitation (WHO and UNISEF, 2000). In general, water pollution has served impacts on the quality of fresh water and aquatic system. Water pollution also has negative impacts on food production, heath and social development and economic activities. Poor quality of surface and groundwater has become a threat to supplies of drinking water throughout the world (World Bank, 1998). In general, the decreasing availability of safe and healthy drinking water due to pollution, in terms of quality and quantity has been a major health concern in Africa and Nigeria in particular.

2.2 Factors Responsible for Water Pollution

There are so many factors which are responsible for water pollution, but it is most often due to human activities. Increasing population, geological factors, rapid urbanization, agricultural developments, global markets, industrial development, industrialization and poor wastewater regulation have affected the quantity and the quality of water (Saleem, 2001; Farooq *et al.*, 2006). Besides the indiscriminate disposal of industrial, municipal and domestic wastes in water channels, rivers, streams and lakes etc. are regarded as the documented source of water pollution (Kahlown and Majeed, 2003).

Asaolu (1998) reported that untreated domestic waste, discharges from industries, rapid deforestation and poor agricultural practices result in the soil erosion and leaching down of nutrients, pesticides and insecticides. An estimated 2 million tons of sewage and other effluents are discharged into the world's waters every day. In developing countries, the situation is worse where over 90% of raw sewage and 70% of untreated industrial wastes are dumped into surface water sources (Asaolu, 1998). Rapid industrialization in urban and Peri-urban areas and high living standards are mainly responsible for discharge of wastewater in the rivers and streams (Ashraf *et al.*, 2010). Other sources of water pollution are sewage and waste water, marine dumping, industrial waste, radioactive waste, oil pollution, underground storage leakages, atmospheric deposition, global warming and eutrophication. The Global Environmental Monitoring System (GEMS) of the United Nations Environmental Program (UNEP) have reported heavy pollution in several rivers around the World (Bichi and Anyata, 1999).

3 Phytoremediation Technology Description

Phytoremediation is a word formed from the Greek prefix "phyto" meaning plant, and the Latin suffix "remedium" meaning to clean or restore (Cunningham *et al.*, 1997). The term actually refers to a diverse collection of plant-based technologies that use either naturally occurring or genetically engineered plants for cleaning contaminated environments. The primary motivation behind the development of phytoremediative technologies is the potential for low-cost remediation (Ensley, 2000). Although the term, phytoremediation, is a relatively recent invention, the practice is not (Brooks, 1998a; Cunningham *et al.*, 1997). Research using semi-aquatic plants for treating radionuclide-contaminated waters existed in Russia at the dawn of the nuclear era (Salt *et al.*, 1995a). Some plants which grow on metalliferous soils have developed the ability to accumulate massive amounts of the indigenous metals in their tissues without exhibiting symptoms of toxicity (Baker and Brooks, 1989; Reeves and Brooks, 1983). Chaney (1983) was the first to suggest using these "hyperaccumulators" for the phytoremediation of metal polluted sites. However, hyperaccumulators were later believed to have limited potential in this area because of their small size and slow growth, which limit the speed of metal removal (Cunningham *et al.*, 1995; Ebbs *et al.*, 1997). By definition, a hyperaccumulator must accumulate at least 1000 μgAg^{-1} of Co, Cu, Cr, Pb, or Ni, or 10,000 μgAg^{-1} (i.e. 1%) of Mn or Zn in the dry matter (Reeves and Baker, 2000; Wantanabe, 1997). Some plants tolerate and accumulate high concentrations of metal in their tissue but not at the level required to be called hyperaccumulators. These plants are often called moderate metal-accumulators, or just moderate accumulators (Kumar *et al.*, 1995). The lack of viable plant alters natives for phytoremediation seemed to suppress the amount of phytoremediation research conducted between the mid 1980s and the early half of the 1990s. The search for plants for phytoremediation centered on the Brassica family to which many hyperaccumulators belong, (Cunningham *et al.*, 1995). Through the work of various researchers, particularly Kumar *et al.* (1995) and Dushenkov *et al.* (1997), several high-biomass, metal-accumulating species were identified. Phytoremediation research gained momentum after the discovery of these plants, and most of our understanding of this emerging technology has come from research reports published since 1995.

Phytoremediation has been increasingly used to clean up contaminated soil and water systems because of its lower costs and fewer negative effects than physical or chemical engineering approaches (Jadia and Fulekar 2008; Reddy and DeBusk, 1986). The principles of phytoremediation system to clean up water bodies include: 1) identification and implementation of efficient aquatic plant systems; 2) uptake of dissolved nutrients including N and P and metals by the growing plants, and the plants creating a favorable environment for a variety of complex chemical, biological and physical processes that contribute to the removal and degradation of nutrients (Gumbricht, 1993); and 3) harvest and beneficial use of the plant biomass produced from the remediation system.

Aquatic plants are utilized for nutrient and metal removal from water because of their fast growth rates, simple growth requirements and ability to accumulate biogenic elements and toxic substances. Since the first recognition of their value in water quality improvement in the 1960s and the 1970s (Sheffield, 1967; Wooten and Dodd, 1976; Asaolu 1998), aquatic plants have been widely used to treat wastewaters or increasingly used to remediate eutrophic waters in forms of constructed wetlands or retention ponds. This is a low-cost treatment with low land requirements, which is attractive to urban areas with high land prices.

Aquatic plants are grouped into submerged, emergent, and floating/floating-leaved aquatic plants according to their leaf's relation with water. Among the submerged aquatic plants, coontail (*Ceratophyllum demersum* L.), hydrilla (*Hydrilla verticillata*), southern naiad (*Najas guadalupensis*) are the most investigated (Lee *et al.*, 1997). Cattail (*Typha latifolia*), bullrush (*Scirpus lacustris*), and common reed (*Phragmites australis*) are the most planted emergent plants in constructed wetlands to remove nutrients such as N and P (Licht, 1990). Among the floating/floating-leaved aquatic plants, water hyacinth (*Eichhornia crassipes*), water lettuce (*Pistia stratiotes*), duckweed (*Lemna spp.* and *Spirodela polyrrhiza* W. Koch), pennywort (*Hydrocotyle umbellata*), and common salvinia (*Salvinia minima* baker) are the best candidates (Licht, 1990; Maine *et al.*, 2004; John *et al.*, 2008; Mishra *et al.*, 2008;). With regard to the uptake capacity of aquatic plants, and subsequently the amount of nutrients or contaminants that can be removed when the biomass is harvested, floating plants (especially large-leaved species) are in the lead, followed by emergent species and then submerged species. Approximately 350 kg P and 2000 kg N ha-1 yr-1 were removed by large-leaved floating plants such as water hyacinths, whereas the capacity of submerged macrophytes was lower (<100 kg P and 700 kg N ha-1 yr-1). Growing in waters with similar P concentrations, water hyacinth had an average P concentration almost twice that of hydrilla, hornwort, pondweed, eelgrass, or naiad, showing a much greater ability for P scavenging (Kidney, 1997). Emergent macrophytes are mostly in the range of 30 to 150 kg P ha-1 yr-1 and 200 to 2500 kg N ha-1 yr-1 (Gumbricht, 1993).

Impressive removal rates of inorganic N (NO_3-N, NH_4-N, and total N) and P (PO_4-P and total P) have been reported from all kinds of phytoremediation systems using aquatic plants especially when invasive floating aquatic plants such as water hyacinth were utilized in nutrient- or metal-rich wastewaters. A wide range of nutrient reduction in wastewaters containing water hyacinth has been reported. For inorganic N, Reddy *et al.* (1982) reported a reduction of about 80%, while Sheffield (1967) observed a 94% reduction. For ortho-P, a 40-55% reduction was reported by Sheffield (1967). For total P, Reddy *et al.* (1982) measured about 32% reduction, while Ornes and Sutton (1975) achieved a much higher removal rate of 80% in their treatment pond. In a pilot scale study using a series of six tanks with water hyacinth for wastewater treatment, the mean decrease in total N and total P in the effluent as it flowed the six tank series was 27.6% and 4.48%, respectively (Licht and Schnoor, 1993). A pond containing water hyacinth, with an air stripping unit and a flocculation and settling unit, was reported to remove >99% ortho-P, 99% nitrate-N, and >99% ammonia-N (Sheffield, 1967). Plant uptake contributes a large proportion to the N and P removal for very high uptake rates have been reported, for instance, 1980 kg N and 322 kg P ha^{-1} y^{-1}, 2500 kg N and 700 kg P ha^{-1} y^{-1} by Rogers and Davis (1972), and up to 5350 kg N ha^{-1} y^{-1} and 1260 kg P ha^{-1} y^{-1} by Reddy and Tucker (1983).

Although at a lower rate compared to such large-leaved floating species as water hyacinth, small-leaved floating species such as duckweed can also remove a considerable amount of nutrients and have been utilized in remediation of wastewaters. Small tank polycultures of duckweed species (*Lemna minor* and *Spirodela polyrhiza*) were found to remove 404 mg N m-2 day-1 (1460 kg N ha-1 yr-1) and 84 mg P m^{-2} day^{-1} (307 kg P ha^{-1} yr^{-1}) from dairy barn wastewater (Komossa *et al.*, 1995). Phosphorus removal rates of 60.0-92.2% were achieved in a wastewater system utilizing *Lemna gibba*. Two species of *Azolla* (*Azolla filiculoides* and *Azolla pinnata*) removed N from mixed waste water resulting in more than 50% decrease in concentration.

According to Ruan *et al.* (2006), polluted river water was efficiently treated by pilot-scale constructed wetland systems planted with emergent aquatic plants, *Typha latifolia* and *Scirpus lacustris*, with mean NH_4-N removal rates of over 85%. Wetlands with emergent macrophytes were reported to remove P at rates from 0.4 to 4.0 g m-2 yr^{-1}, with more eutrophic systems achieving higher removal rate (Komossa *et al.*, 1995).

Lu (2009) observed an increase in transparency and a decrease in the concentrations of P simultaneously with increased presence of submerged macrophytes in the lake.

Aquatic plants also demonstrate tremendous potential in metal accumulation and removal from the surrounding waters. Free water surface and subsurface flow pilot-size wetlands were constructed to treat highway runoff with metal removal rates of 47%, 23%, 33%, and 61% for Cu, Ni, Pb and Zn, respectively, with their respective two-year mean concentrations of 56, 114, 49 and 250 μg L-1 (Madison, 1998). *Azolla filiculoides* removed 91.0, 41.5, 82.5, 37.7, 12.1, 46.7 and 67.2% of the initial Fe, Zn, Cu, Mn, Co, Cd and Ni, respectively from mixture of waste waters, while *Azolla pinnata* removed 92.7, 83.0, 59.1, 65.1, 95.0, 90.0 and 73.1%, respectively. Although all three plants, water lettuce (*Pistia stratiotes* L.), duckweed (*Spirodela polyrrhiza* W. Koch), and water hyacinth (*Eichhornia crassipes*) demonstrated high removal rates of Fe, Zn, Cu, Cr, and Cd (>90%) without reduction in growth, water hyacinth were the most efficient followed by water lettuce and duckweed (Mishra and Tripathi, 2008). Many researchers have reported that high heavy metal concentrations (Cu, Cd, Mn, Pb, Hg, etc.) were measured in the tissues of aquatic plant growing in waters with elevated metal concentrations and no toxic effects or reduction in plant growth were observed (Licht, 1997; Mishra *et al.*, 2008; Ugya et al., 2015d).

Common duckweed and water hyacinth have been reported to be the top species as Cd accumulators (Wang *et al.*, 2002; Zayed *et al.*, 1998; Zhu *et al.*, 1999; Ugya, 2015; Ugya and Imam, 2015). Both *Salvinia herzogii* and *Pistia stratiotes* efficiently removed Cr from water at the concentrations of 1, 2, 4, and 6 mgL^{-1} (Maine *et al.*, 2004). Lead concentrations in plant tissue (mg kg-1) were found to be 1621 and 1327 times those in the external solution (mg L^{-1})

for *C. demersum* and *C. caroliniana*, respectively (Fonkou *et al.*, 2005). *Salvinia minima* has been reported as a hyperaccumulator of Cd and Pb with bioconcentration factors (metal concentration in plant tissue over that in external solution) of approximately 3000 for both heavy metals (Clough *et al.*, 1987; Ugya et al., 2015c)

4 Growth Factors of Aquatic Plants

For a phytoremediation system to work efficiently, optimal plant growth is the key. Many environmental factors can influence plant growth and its performance, such as temperature, nutrient concentration, pH, solar radiation, and salinity of the water. The weight and size of aquatic plants are a function of these factors. For example, growth of water hyacinth plants cultured in nutrient solution were significantly influenced by the seasonal changes in temperature and solar radiation, shorter time was required to reach maximum biomass yield in summer with high growth rate (Reddy *et al.*, 1983; Ugya et al., 2015a; Ugya et al., 2015b). If maximum growth is obtained, one hectare of water hyacinths could remove about 2500 kg N yr^{-1} (Rogers and Davis, 1972) and as high as 7629 kg N ha-1 yr^{-1} was reported by Reddy and Tucker (1983) for water hyacinth cultured in a nutrient solution.

Although large-leaved floating plants such as water hyacinth and water lettuce can produce high biomass and remove large amounts of nutrients and metals, they may not be suitable for temperate or frigid areas due to their sensitivity to cool temperature which significantly affects their performance (Clough *et al.*, 1987). Instead, duckweed or azolla could be a better choice because of their tolerance to colder weather (Reddy *et al.*, 1983). This also explains why pennywort removed 20% more N and 30% more P from primary domestic effluent than water hyacinth during the winter in central Florida (Clough *et al.*, 1987).

Nutrient availability affects the growth and performance of aquatic plants. Within the studied nutrient concentration ranges, mean number of ramets, mean height and total biomass of water hyacinth significantly increased with increasing nutrient level (Zhao *et al.*, 2006). A 200-fold difference in dry weight of water lettuce was reported by Aoi and Hayashi (1996) between cultivated in rain water and treated sewage water. Similar to terrestrial species, aquatic plants respond positively to nutrient concentration increases up to a certain point followed by no further response or a negative response. Five and a half mg per liter and 1.06 mgL^{-1} were such points reported for water hyacinth growth, while 20 mgL^{-1} and 2 mgL^{-1} were found for *Salvinia molesta* (Cary and Weerts, 1984). Not only nutrient concentration itself, but also ratios between different nutrients play an important role in plant growth. It was reported that the highest production of water hyacinth occurs when the N:P ratio in the water was close to 3.6 (Reddy and Tucker, 1983).

Most water bodies varied in salinity which may have significant effects on aquatic plants' growth and performance. Utilization of such invasive aquatic plants as water hyacinth and water lettuce has its advantages as discussed above and its concern of plant escape from the detention systems into the lagoons or estuaries. Knowledge on salinity tolerance of candidate plant(s) can help better utilize the plant(s) without bringing disaster. Salt concentrations of 1660 and 2500 mg kg-1 (equivalent to 2683 and 4040 μS cm^{-1}) were reported to have toxic effects on water lettuce and water hyacinth, respectively (Haller *et al.*, 1974).

pH plays a role in plant growth directly by hydrogen ion (H$^+$) injury at low pH and indirectly by affecting availability and toxicity of mineral elements (Mahmood *et al.*, 2005). Generally, plant grows best in the pH range of 5.5-7.0. Optimum pH ranges 6.5-7.5 and 5.8-6.0 were reported for water hyacinth (El-Gendy *et al.*, 2004; Hao and Shen, 2006). Macroalgae *Chlorella sorokiniana* grew best at pH 7-8 (Moronta *et al.*, 2006).

5 Phytoremediation Case Studies

Phytoremediation consists of a collection of four different plant-based technologies, each having a different mechanism of action for the remediation of metal-polluted soil, sediment, or water. These include: rhizofiltration, which involves the use of plants to clean various aquatic environments; phytostabilization, where plan ts are used to stabilize rather than clean contaminated soil; phytovolatilization, which involves the use of plants to extract certain metals from soil and then release them into the atmosphere through volatilization; and phytoextraction, where plants absorb metals from soil and translocate them to the harvestable shoots where they accumulate. Although plants show some ability to reduce the hazards of organic pollutants, the greatest progress in phytoremediation has been made with metals (Salt *et al.*, 1995; Watanabe, 1997; Blaylock and Huang, 2000). Phytoremediative technologies which are soil-focused are suitable for large areas that have been contaminated with low to moderate levels of contaminants.

Sites which are heavily contaminated cannot be cleaned through phytoremediative means because the harsh conditions will not support plant growth. The depth of soil which can be cleaned or stabilized is restricted to the root zone of the plants being used. Depending on the plant, this depth can range from a few inches to several meters (Schnoor *et al.*, 1997). Phytoremediation should be viewed as a long-term remediation solution because many cropping cycles may be needed over several years to reduce metals to acceptable regulatory levels. This new remediation technology is competitive with, and may be superior to existing conventional technologies at sites where phytoremediation is applicable. Phytoremediation is not the solution for all hazardous waste problems but is rather a tool that can be used, possibly in conjunction with other clean-up methods, to remediate polluted environments.

5.1 Phytoextraction

This technology involves the extraction of metals by plant roots and the translocation thereof to shoots. The roots and shoots are subsequently harvested to remove the contaminants from the soil. Salt et al. (1995) reported that the costs involved in phytoextraction would be more than ten times less per hectare compared to conventional soil remediation techniques. Phytoextraction also has environmental benefits because it is considered a low impact technology. Furthermore, during the phytoextraction procedure, plants cover the soil and erosion and leaching will thus be reduced. With successive cropping and harvesting, the levels of contaminants in the soil can be reduced (Vandenhove *et al.*, 2001).

Researchers at the University of Florida have discovered the ability of the Chinese brake fern, P. vittata to hyperaccumulate arsenic. In a field test, the ferns were planted at a wood-preserving site containing soil contaminated with from 18.8 to 1,603 parts per million arsenic, and they accumulated from 3,280 to 4,980 parts per million arsenic in their tissues (Ma *et al.*, 2001). Sunflower, H. annus have proven effective in the remediation of radionuclides and certain other heavy metals. The flowers were planted as a demonstration of phytoremediation in a pond contaminated with radioactive cesium- 137 and strontium-90 as a result of the Chernobyl nuclear disaster in the Ukraine. The concentration of radionuclides in the water decreased by 90% in a two week period. According to the demonstration, the radionuclide concentration in the roots was 8000 times than that in the water. In a demonstration study performed by Phytotech for the Department of Energy, H. annus reduced the uranium concentration at the site from 350 parts per billion to 5 parts per billion, achieving a 95% reduction in 24 h (Schnoor, 1997).

5.2 Phytostabilization

Phytostabilization, also referred to as in-place inactivation, is primarily used for the remediation of soil, sediment, and sludges (United States Protection Agency, 2000). It is the use of plant roots to limit contaminant mobility and bioavailability in the soil. The plants primary purposes are to (1) decrease the amount of water percolating through the soil matrix, which may result in the formation of a hazardous leachate, (2) act as a barrier to prevent direct contact with the contaminated soil and (3) prevent soil erosion and the distribution of the toxic metal to other areas. Phytostabilization can occur through the sorption, precipitation, complexation, or metal valence reduction. It is useful for the treatment of lead (Pb) as well as arsenic (As), cadmium (Cd), chromium (Cr), copper (Cu) and zinc (Zn).

Some of the advantages associated with this technology are that the disposal of hazardous material/biomass is not required (United States Protection Agency, 2000) and it is very effective when rapid immobilization is needed to preserve ground and surface waters. The presence of plants also reduces soil erosion and decreases the amount of water available in the system (United States Protection Agency, 2000). Phytostabilization has been used to treat contaminated land areas affected by mining activities and Superfund sites. The experiment on phytostabilization by Jadia and Fulekar (2008) was conducted in a greenhouse, using sorghum (fibrous root grass) to remediate soil contaminated by heavy metals and the developed vermicompost was amended in contaminated soil as a natural fertilizer. They reported that growth was adversely affected by heavy metals at the higher concentration of 40 and 50 ppm, while lower concentrations (5 to 20 ppm) stimulated shoot growth and increased plant biomass. Further, heavy metals were efficiently taken up mainly by roots of sorghum plant at all the evaluated concentrations of 5, 10, 20, 40 and 50 ppm. The order of uptake of heavy metals was: Zn>Cu>Cd>Ni>Pb. The large surface area of fibrous roots of sorghum and intensive penetration of roots into the soil reduces leaching via stabilization of soil and capable of immobilizing and concentrating heavy metals in the roots.

5.3 Rhizofiltration

Rhizofiltration is primarily used to remediate extracted groundwater, surface water, and wastewater with low contaminant concentrations (Ensley, 2000). It is defined as the use of plants, both terrestrial and aquatic, to absorb, concentrate, and precipitate contaminants from polluted aqueous sources in their roots. Rhizofiltration can be used for Pb, Cd, Cu, Ni, Zn, and Cr, which are primarily retained within the roots (United States Protection Agency, 2000). Sunflower, Indian mustard, tobacco, rye, spinach, and corn have been studied for their ability to remove lead from water, with sunflower having the greatest ability. Indian mustard has a bioaccumulation coefficient of 563 for lead and has also proven to be effective in removing a wide concentration range of lead (4 mg/L -500 mg/L) (United States Protection Agency, 2000). The advantages associated with rhizofiltration are the ability to use both terrestrial and aquatic plants for either in situ or ex situ applications. Another advantage is that contaminants do not have to be translocated to the shoots. Thus, species other than hyperaccumulators may be used. Terrestrial plants are preferred because they have a fibrous and much longer root system, increasing the amount of root area. Sunflower (*Asteracaea spp.*) have successfully been implemented for rhizofiltration at Chernobyl to remediate uranium contamination. Dushenkov *et al.* (1997) observed that roots of many hydroponically grown terrestrialplants s uch as Indian mustard (*B. juncea* (L.) Czem) and sunflower (*H. annuus* L.) effectively removed the potentially toxic metals, Cu, Cd, Cr, Ni, Pb and Zn, from aqueous solutions.

An experiment on rhizofilteration by Karkhanis *et al.* (2005) was conducted in a greenhouse, uisng pistia, duckweed and water hyacinth (*Eichornia* crassipes) to remediate aquatic environment contaminated by coal ash containing heavy metals. Rhizofilteration of coal ash starting from 0, 5, 10, 20, 30, 40%. Simultaneously the physicochemical parameters of leachate have been analyzed and studied to understand the leachability. The results showed that pistia has high potential capacity of uptake of the heavy metals (Zn, Cr, and Cu) and duckweed also showed good potential for uptake of these metals next to pistia. Rhizofiltration of Zn and Cu in case of water hyacinth was lower as compared to pistia and duckweed. This research shows that pistia/duckweed/water hyacinth can be good accumulators of heavy metals in aquatic environment.

5.4 Phytovolatilization

Phytovolatilization involves the use of plants to take up contaminants from the soil, transforming them into volatile forms and transpiring them into the atmosphere (United States Protection Agency, 2000). Mercuric mercury is the primary metal contaminant that this process has been used for. The advantage of this method is that the contaminant, mercuric ion, may be transformed into a less toxic substance (that is, elemental Hg). The disadvantage to this is that the mercury released into the atmosphere is likely to be recycled by precipitation and then redeposited back into lakes and oceans, repeating the production of methyl-mercury by anaerobic bacteria.

In laboratory experiments, tobacco (*N. tabacum*) and a small model plant (*Arabidopsis thaliana*) that had been genetically modified to include a gene for mercuric reductase converted ionic mercury (Hg(II)) to the less toxic metallic mercury (Hg(0)) and volatilized it (Meagher *et al.*, 2000). Similarly transformed yellow poplar (*Liriodendron tulipifera*) plantlets had resistance to, and grew well in, normally toxic concentrations of ionic mercury. The transformed plantlets volatilized about ten times more elemental mercury than did untransformed plantlets. Indian mustard and canola (*Brassica napus*) may be effective for phytovolatilization of selenium, and, in addition, accumulate the selenium (Bañuelos *et al.*, 1997).

5.5 Plant-Metal Uptake

Plants extract and accumulate metals from soil solution. Before the metal can move from the soil solution into the plant, it must pass the surface of the root. This can either be a passive process, with metal ions moving through then porous cell wall of the root cells, or an active process by which metal ions move symplastically through the cells of the root. This latter process requires that the metal ions traverse the plasmalemma, a selectively permeable barrier that surrounds cells (Pilon-Smits, 2005). Special plant membrane proteins recognize the chemical structure of essential metals; these proteins bind the metals and are then ready for uptake and transport. Numerous protein transporters exist in plants. For example, the model plant thale cress (*A. thaliana*) contains 150 different cation transporters (Axelsen and Palmgren, 2001) and even more than one transporter for some metals. Some of the essential, nonessential and toxic metals, however, are analogous in chemical structure so that these proteins regard them as the same. For example arsenate is taken up by P transporters. Abedin *et al.* (2002) studied the uptake kinetics of as species, arsenite and arsenate, in rice plants and found that arsenate uptake was strongly suppressed in the presence of arsenite. Clarkson and Luttge (1989) reported that Cu and Zn, Ni and Cd compete for the same membrane carriers. For root to shoot transport these elements are transported via the vascular system to the above-soil biomass (shoots). The shoots are harvested, incinerated to reduce volume disposed of as hazardous waste, or precious metals can be recycled (phytomining). Different chelators may be involved in the translocation of metal cations through the xylem, such as organic acid chelators [malate, citrate, histidine (Salt *et al.*, 1995; von Wiren *et al.*, 1999), or nicotianamine (Stephen *et al.*, 1996; von Wiren *et al.*, 1999)]. Since the metal is complexed within a chelate it can be translocated upwards in the xylem without being adsorbed by the high cation exchange capacity of the xylem (von Wiren *et al.*, 1999).

6 Role of Macrophytes in Water Contamination Removal

Macrophytes play important roles in balancing Lake Ecosystem. For the first time, they were recognised during 1960s and 1970s in water quality improvement (Wooten and Dodd, 1976). Aquatic macrophytes treatment systems for waste-water are the need of developing countries, because they are cheaper to construct and a little skill is required to operate (Mahmood *et al.*, 2005). They improve the water quality by absorbing nutrients with their effective root system (Dhote and Dixit, 2007). Macrophytes not only retain nutrients by biomass uptake, but also increases sedimentation (Schulz *et al.*, 2003). These are utilized for nutrient and metal removal from water in the forms of CW or retention ponds because of their fast growth rates, simple requirements, and ability to accumulate biogenic elements and toxic substances (Lu, 2009). Aquatic plants are grouped into submerged, emergent, and floating-leaved based on their leaf's relation with water. During selection, biomass production, growth rate, and easiness of management and harvest should be taken into account (Lu, 2009). Wetlands are mainly dominated by the floating aquatic macrophytes (DeBusk and Reddy, 1987; Brix and Shierup, 1989; Vymazal *et al.*, 1998). Floating aquatic plants can grow in vertical as well as horizontal direction, thereby increasing the photosynthetic surface area. These factors altogether makes floating aquatic plants, one of the earth's most productive communities (Lu, 2009). The most common aquatic macrophytes among the floating-leaved, being employed in wastewater treatment are water hyacinth, water lettuce and Duckweed (John *et al.*, 2008; Maine *et al.*, 2004; Mishra *et al.*, 2008). Impressive removal rates of inorganic nitrogen [nitrate (NO_3-N), ammonium (NH_4-N), and total N)] and phosphorus (PO4-P and total P) have been reported using aquatic plants especially when water hyacinth were utilized in nutrient or metal-rich wastewaters (Lu, 2009). Awuah *et al.* (2004) found 70% of TDS reductions by water lettuce.

6.1 Phytoremediation of Toxic Elements by Aquatic Macrophytes

Freshwater as well as seawater resources are being contaminated by various toxic elements through anthropogenic activities and from natural sources. Therefore, remediation of contaminated aquatic environment is important as it is for terrestrial environment. Phytoremediation of the toxic contaminants can be readily achieved by aquatic macrophytes or by other floating plants since the process involves biosorption and bioaccumulation of the soluble and bioavailable contaminants from water (Brooks et al., 1998). In aquatic phytoremediation systems, aquatic plants can be either floating on the water surface or submerged into the water. The floating aquatic hyperaccumulating plants absorb or accumulate contaminants by its roots while the submerged plants accumulate metals by their whole body. Many years ago, Hutchinson (1975) reviewed the ability of aquatic macrophytes to concentrate elements from the aquatic environment and reported that the levels of potentially toxic elements in the plants were at least an order of magnitude higher than in the supporting aqueous medium. Later on, Brooks et al., (1998) reviewed the hyperaccumulation of toxic trace elements by aquatic vascular plants and discussed about the pathways and rates of elemental uptake and excretion, environmental factors that control uptake of elements, and the significance of trace elements uptake for the field of wastewater treatment and biomonitoring of pollutants, which is of great interest for bioremediation of aquatic systems. By this time, considerable number of literatures have been published which described different aspects of biogeochemistry, mechanisms and uptake of toxic elements by a large number of aquatic macrophytes to develop effective phytoremediation technology.

Several aquatic macrophytes and some other small aquatic floating plants have been investigated for the remediation of natural and wastewater contaminated with Cu(II), Cd(II) and Hg(II) (Sen and Mondal, 1987; 1991; Alam *et al.*, 1995).

Microspora and *Lemna minor* were studied for Pb and Ni remediation (Axtell *et al.*, 2003). Five common aquatic plant species (*Typha latifolia, Myriophyllum exalbescens, Potamogeton epihydrus, Sparganium angustifolium* and *Sparganium multipedunculatum*) were tested for Al phytoremediation (Alam et al., 1995). Parrot feather (*Myriophyllum aquaticum)*, creeping primrose (*Ludwigina palustris*), and water mint (Mentha aquatic) have been reported to remove Fe, Zn, Cu, and Hg from contaminated water effectively (Kara, 2004). The L. minor was reported to accumulate Cu and Cd from contaminated wastewater (Kara, 2004; Hou *et al.*, 2007). The submerged aquatic plant *Myriophyllum spicatum* L. has been reported as an efficient plant species for the metal-contaminated industrial wastewater treatment (Hou *et al.*, 2007). The aquatic plants *Rorippa nasturtium-aquaticum* (L.) and *Mentha spp* accumulate arsenic from contaminated freshwater (Kara, 2004).

Conclusion

Effluents can have ecological impact on water bodies leading to increased nutrient load especially if they are essential metals. These metals in effluent may increase fertility of water leading to euthrophication, which in open water can progressively lead to oxygen deficiency, algae blooms and death of aquatic life (Pickering and Owen, 1997). Heavy metals can bioaccumulate and through the food chain, to toxic level in man. Mercury can cause numbness, locomotory disorder, brain damage, convulsion and nervous problems. Cadmium is responsible for kidney tubular impairment and osteomalacia. Cadmium, zinc and manganese are reported to affect ion balance if present in sufficient amount. This study shows that phytoremediation has a great role to play in the remediation of industrial waste water.

References

Abedin, M.J., Feldmann, J. and Meharg A.A. (2002); Uptake Kinetics of Arsenic Species in Rice Plants.*Plant Physiol.* 128: 1120-28.

Aghalino, S. O. and Eyinla, B. (2009); Oil Exploration and Marine Pollution: Evidence from the Niger Delta, Nigeria", *Journal of Human Ecology*, 28 (3), 177-82.

Akpor, O.B. and Muchie, M. (2010): Remediation of Heavy Metals in Drinking Water and Wastewater Treatment Systems: Processes and Applications. *Int. J. Phys. Sci.* 5(12):1807-1817.

Alam, B., Chatterjee, A.K. and Dattagupta, S. (1995); Bio Accumulation of Cd (I) By Water Lettuce (*Pistia stratioles* L.).,*Poll. Research*, 14 (1), 59-64.

Aoi, T. and Hayashi, T. (1996); Nutrient Removal By Water Lettuce (*Pistia stratiotes*).,*Water Sci. Technol.*, 34(7-8): 407-412.

Asaolu, S.S. (1998); Chemical Pollution Studies of Coastal Water of Ondo State. Ph.D Thesis, Fed. Univ. Technol., Akure. (Unpublished).

Ashraf, M. A., Maah, M. J., Yusoff, I. and Mehmood, K. (2010); Effects of Polluted Water Irrigation on Environment and Health of People in Jamber, District Kasur, Pakistan, *International Journal of Basic & Applied Sciences,* 10(3), pp. 37-57.

Axtell, N. R., Sternberg, S. P. K., and Claussen, K. (2003); Lead and Nickel Removal Using Microspora and *Lemna minor.Bioresource Technology*, 89, 41-48.

Axelsen, K.B. and Palmgren, M.G. (2001); Inventory of the Superfamily of P-type Ion Pumps in Arabidopsis.*Plant Physiol.* 126: 696-706.

Awuah, E., Oppong-Peprah, M., Lubberding, H.J. and Gijzen, H.J. (2004); Comparative Performance Studies of Water Lettuce, Duckweed and Algal-Based Stabilization Ponds Using Low-Strength Sewage.,*J. Toxicol. Environ. Health-Part A.*, 67(20-22), 1727-1739.

Azizullah, A., Khattak, M.N., Richter P. qand Hader, D.P. (2011): Water Pollution in Pakistan and its Impact on Public Health- A review. *Environ. Int.* 37(02):479-97.

Bañuelos, G.S., Ajwa, H.A., Mackey, B., Wu L.L., Cook, C., Akohoue S. and Zambrzuski S (1997); Evaluation of Different Plant Species Used For Phytoremediation of High Soil Selenium.*J. Environ.* Qual. 26(3):639- 646.

Baskar, G., Deeptha, V.T. and Rahman, A.A. (2009); Treatment of Wastewater from Kitchen in an Institution Hostel Mess Using Constructed Wetland.,*Int. J. Recent Trends Eng.*, 1(6), 54-58.

Bichi, M.H. and Anyata, B.U. (1999); Industrial waste pollution in the Kano River Basin. Environ. Manag. Health 10:112-116.

Blaylock, M. and Huang, J. (2000); Phytoextraction of metals. In: Raskin I, Ensley B (eds) Phytoremediation of Toxic Metals: Using Plants to Clean Up the Environment. Wiley, New York, pp 53–69

Brix, H. and Shierup, H.H. (1989); The Use of Aquatic Macrophytes in Water Pollution Control., Ambio, 18, 100-107.

Brooks, R., Chambers, M., Nicks, L. and Robinson B. (1998); Phytomining. *Trends Plant Sci* 3(9):359–362

Brooks, R.R. (1998); Plants that Hyperaccumulate Heavy Metals, their Role in Phytoremediation, Microbiology, Archeology, Mineral Exploration and Phytomining. CAB Internat, New York.

Cary, P.R., and Weerts, P.G.J. (1984); Growth of *Salvinia molesta* as Affected By Water Temperature and Nutrition. III. Nitrogen-Phosphorus Interactions and Effect of pH. *Aquat. Bot.* 19:171-182.

Chaney, R. L. (1983); Plant Uptake of Inorganic Waste Constituents. pp. 50-76. In J. F. Parr, P. B. Marsh, and J. M. Kla (eds.), Land Treatment of Hazardous Waste. Noyes Data Corporation, Park Ridge, NJ.

Clarkson D.T. and Luttge U. (1989); Mineral Nutrition: Divalent Cations, Transport and Compartmentation. Progr. Bot. 51: 93-112.

Clough, .K.S. , DeBusk, T.A. and Reddy, K.R. (1987); Model Water Hyacinth and Pennywort Systems for the Secondary Treatment of Domestic Wastewater., In: Aquatic Plants for Water Treatment and Resource Recovery, K.R. Reddy and W.H. Smith (ed.), Magnolia Publishing Inc., Orlando, FL, pp. 1031.

Cunningham, S. D., J. R. Shann, D. E. Crowley, and T. A. Anderson.(1997); Phytoremediation of Contaminated Water and Soil. In E. L. Kruger, T. A. Anderson, and J.

R. Coats (eds.), Phytoremediation of Soil and Water Contaminants, ACS Symposium Series No. 664. American Chemical Society, Washington, DC.

Cunningham, S. April 19-22, (1995); In Proceedings/Abstracts of the Fourteenth Annual Symposium, Current Topics in Plant Biochemistry, – Physiology, and Molecular Biology Columbia; pp. 47-48.

Dar, S.H., Kumawat, D.M., Singh, N. and Wani, K.A. (2011); Sewage Treatment Potential of Water Hyacinth (*Eichhornia crassipes*).,*Res. J. Environ. Sci.*, 5(4), 377-385.

DeBusk, .T.A. and Reddy, K.R. (1987); Wastewater Treatment Using Aquatic Macrophytes: Contaminant Removal Processes and Management Strategies., In: Aquatic Plants for Water Treatment and Resource Recovery, K.R. Reddy and W.H. Smith (eds.), Mongolia publishing Inc., Orlando.

Dhote, S. and Dixit, S. (2007); Water Quality Improvement Through Macrophytes: A Case Study.,*Asian J. Exp. Sci.*, 21(2), 427-430.

Dipu, S., Kumar, A.A and Thanga, V.S.G. (2011); Phytoremediation of Dairy Effluent by Constructed Wetland Technology.,*Environmentalist*, 31, 263-278.

Dushenkov, S., Vasudev, D., Kapulnik, Y., Gleba, D., Fleisher, D., Ting, K. C. and Ensley, B. D. (1997); Removal of uranium from water using terrestrial plants.*Environmental Science and Technology* **31**: 3468-3474.

Ebbs, S. D., M. M., Lasat, D. J., Brady, J., Cornish, R., Gordon, L.V.(1997); Phytoextraction of Cadmium and Zinc from a Contaminated Soil.*J. Environ. Qual.* 26(5):1424-1430.

El-Gendy, A.S., Biswas, N. and Bewtra, J.K. (2004); Growth of Water Hyacinth in Municipal Landfill Leachate with Different pH., *Environ. Technol.*, 25, 833-840.

Ensley, B.D. (2000) "Rationale for the Use of Phytoremediation." Phytoremediation of Toxic Metals:Using Plants to Clean-Up the Environment. John Wiley Publishers: New York.

Farooq, H., Siddiqui, M.T., Farooq, M., Qadir, E. and Hussain Z (2006); Growth, Nutrient Homeostatis and Heavy Metal Accumulation in *Azadirachta indica* and *Dalbrgia sissoo* Seedlings Raised from Waste Water.Int.

Fonkou, T., Agendia, P., Kengne, I., Akoa, A. and Nya, J. (2002); Potentials of Water Lettuce (*Pistia stratiotes*) in Domestic Sewage Treatment with Macrophytic Lagoon Systems in

Cameroon., In: Proc. of International Symposium on Environmental Pollution Control and Waste Management, Tunis, 709-714.

Foster, S.S.D. and Hirata. R. (1988); *Groundwater Pollution Risk Assessment: A Method Using Available Data*. Pan American Centre for Sanitary Engineering and Environmental Science (CEPIS), Lima.

Foster, S.S.D. and Gomes, D.C. (1989); *Groundwater Quality Monitoring: An Appraisal of Practices and Costs*. Pan American Centre for Sanitary Engineering and Environmental Science (CEPIS), Lima

GESAMP (1988); *Report of the Eighteenth Session, Paris 11-15 April 1988*. GESAMP Reports and Studies No. 33, United Nations Educational, Scientific and Cultural Organization, Paris.

Girija, N., Pillai, S.S and Koshy, M. (2011); Potential of Vetiver for Phytoremediation of Waste in Retting Area., The Ecoscan, 1, 267-273.

Gumbricht, T. (1993); Nutrient removal processes in freshwater submersed macrophyte systems., Ecol. Eng., 2, 1-30.

Haller, W.T., Sutton, D.L. and Barlowe, W.C. (1974); Effects of Salinity on Growth of Several Aquatic Macrophytes., Ecology, 55, 891-894

Hao, F.L., and M.W. Shen.(2006); Effect of pH Adjustment on Purifying Efficiency of Water Hyacinth in Cultivated Wastewater.*Journal of Shanghai Jiaotong University - Agricultural Science* 24:196-199.

Hou,W., Chen, X., Song, G., Wang, Q., and Chang, C. C. (2007); Effects of Copper and Cadmium on Heavy Metal Polluted Water body Restoration by Duckweed (*Lemna minor*) *Physiology. Plant Biochemistry*, 45, 62–69.

Hutchinson G.E. (1975); A Treatise or Limnology: Vol 3; Limnological Botany. New York. John Wiley and Sons.

Hem, J.D. (1984); *Study and Interpretation of the Chemical Characteristics of Natural Water.3rd edition. Water Supply Paper 2254, United States Geological Survey, Washington, DC*

Jadia, C.D. and Fulekar M.H. (2008); Phytotoxicity and Remediation of Heavy Metals by Fibrous Root Grass (sorghum).*Journal of Applied Biosciences*. (10): 491 - 499. ISSN 1997 – 5902: www.biosciences.elewa.org

Jamuna, S. and Noorjahan, C.M. (2009); Treatment of Sewage Waste Water Using Water Hyacinth - *Eichhornia sp* and Its Reuse for Fish Culture.,*Toxicol. Int.*, 16(2), 103-106.

Jing, S.R., Lin, Y.F., Wang, T.W. and Lee, D.Y. (2002); Microcosm Wetlands for Wastewater Treatment with Different Hydraulic Loading Rates and Macrophytes.,*J. Environ. Qual.*, 31, 690-696.

John, R., Ahmad, P., Gadgil, K. and Sharma, S. (2008); Effect of Cadmium and Lead on Growth, Biochemical Parameters and Uptake in *Lemna polyrrhiza* L., *Plant Soil Environ.*, 54, 262-270.

Kara, Y., Basaran, D., Kara, I., Zeytunluoglu, A., and Genc, H. (2003); Bioaccumulation of Nickel by Aquatic Macrophyta *Lemna minor* (Duckweed).*International Journal of Agriculture and Biology*, 5 (3), 281–283.

Kahlown, M.A. and Majeed, A. (2003); Water-resources situation in Pakistan: challenges and future strategies. Water resources in the south: Present scenario and future prospects in Islamabad, Pakistan. Comm. Sci. Technol. Sustain. Dev. South (COMSATS) pp. 21–39.

Karkhanis, M., Jadia, C.D. and Fulekar, M.H. (2005).Rhizofilteration of Metals from Coal Ash Leachate. *Asian J. Water, Environ. Pollut* 3(1): 91-94.

Kidney, S. (1997); Phytoremediation May Take Root in Brownfields. In: The Brownfields Report, Volume 2, No. 14, pp. 1, 6, 7.

Komossa, D., Langebartels, C., and Sandermann, Jr. H (1995); Metabolic Processes for Organic Chemicals in Plants. In S. Trapp and J. C. McFarlane (eds.), Plant Contamination: Modeling and Simulation of Organic Chemical Processes. Lewis Publishers, Boca Raton, FL.

Kumar, P. B. A. N., Dushenkov, V., Motto, H. and Raskin, I. (1995); 1. Phytoextraction: the Use of Plants to Remove Heavy Metals form Soils. *Environmental Science and Technology* **29**: 1232-1238.

Lakshmana, P.P., Jayashree, S. and Rathinamala, J. (2008); Application of Vetiver for Water and Soil Restoration., Available Online: http://www.vetiver.org/TVN/India 1st workshop proceeding//Chapter2-3 pdf.

Lee, R.W., S.A. Jones, E.L. Kuniansky, G.J. Harvey, and S.M. Eberts.(1997); Phreatophyte Influence On Reductive Dechlorination In a Shallow Aquifer Containing TCE. pp. 263-268. In G.B. Wickramanayake and R.E. Hinchee (eds.) Bioremediation and Phytoremediation, Chlorinated and Recalcitrant Compounds.Battelle Press, Columbus, OH.

Lu Q. (2009); Evaluation of aquatic plants for phytoremediation of eutrophic stormwaters.,Ph.D Thesis, University of Florida, Florida.

Lu, Q., He, Z.L., Graetz, D.A., Stoffella, P.J. and Yang, X. (2010); Phytoremediation to Remove Nutrients and Improve Eutrophic Stormwaters Using Water Lettuce (*Pistia stratiotes* L.).,*Environ. Sci. Poll. Res.*, 17, 84-96.

Lissy, A.M.P.N, and Madhu, B.Dr.G. (2010); Removal of Heavy Metals from Waste Water Using Water Hyacinth., In: Proc. of the International Conference on Advances in Civil Engineering, 42-47.

Licht, L. A. (1990); Poplar Tree Buffer Strips Grown in Riparian Zones for Biomass Production and Nonpoint Source Pollution Control. Ph.D. Thesis, University of Iowa, Iowa City, IA.

Licht, L. A., and Schnoor, J.L. (1993); Tree Buffers Protect Shallow Ground Water at Contaminated Sites. EPA Ground Water Currents, Office of Solid Waste and Emergency Response.EPA/542/N-93/011.

Licht, L. A. (1997); A Retrospective-10 Growing Seasons Using Populus Spp. for Pollution Control. 12th Annual Conf.On Hazardous Waste Research, May 19-22, 1997, Kansas City, MO.

Ma, L.Q., Komar, K.M., Tu, C., Zhang, W., Cai, Y.and Kennelley, E.D. (2001); A Fern That Hyperaccumulates Arsenic. Nature; 409: 579-579.

Madison, M. F., Litch, L.A., Ricks, F.B. and Grabhorn, H. (1991); Landfill Cap Closure Utilizing A Tree Ecosystem.The American Society of Agricultural Engineers. June 23-26. Albuquerque, NM. Madison, M. F. 1998. Personal Communication. March 1998.

Maine, M.A., Sune, N.L. and Lagger, S.C. (2004); Chromium Bioaccumulation: Comparison of the Capacity of Two Floating Aquatic Macrophytes.,*Water Res.*, 38, 1494-1501.

Mattina MJI, Berger WL, Musante C, White C (2003); Concurrent plant uptake of heavy metals and persistent organic pollutants from soil. *Environ. Pollut.* 124:375-378.

Mahmood, Q., Zheng, P., Islam, E., Hayat, y., Hassan, M.J., Jilani, G. and Jin, R.C. (2005); Lab Scale Studies on Water Hyacinth (*Eicchornia crassipes* mart solms) for Biotreatment of Textile Waste Water.*Caspian J. Env.Sci.*, 3(2): 83-88.

Mashauri, D.A., Mulungu, D.M.M. and Abdulhussein, B.S. (2000); Constructed Wetland at the University of Dar es salaam., Wat. Res., 34(4), 1135-1144.

McJunkin, F.E. (1982); *Water and Human Health.*United States Agency for International Development, Washington, DC.

Meybeck, M. and Helmer, R. (1996); Introduction. In: D. Chapman [Ed.] *Water Quality Assessments. A Guide to the Use of Biota, Sediments and Water in Environmental Monitoring.*2nd edition.Chapman & Hall, London.

Meagher RB, Rugh CL, Kandasamy MK, Gragson G, Wang NJ (2000); Engineered Phytoremediation of Mercury Pollution in Soil and Water Using Bacterial Genes. In: Terry N, Bañuelos G (eds) Phytoremediation of Contaminated Soil and Water. Lewis Publishers, Boca Raton, FL. pp. 201-219.

Mishra, V.K. and Tripathi, B.D. (2008); Concurrent Removal and Accumulation of Heavy Metals by the Three Aquatic Macrophytes.,*Bioresource Technol.*, 99, 7091-7097.

Mishra, V.K., Upadhyay, A.R., Pandey, S.K. and Tripathi, B.D.(2008); Heavy Metal Pollution Induced Due to Coal Mining Effluent On Surrounding Aquatic Ecosystem And Its Management Through Naturally Occurring Aquatic Macrophytes., *Bioresource Technol.*, 99, 930-936.

Mkandawire, M. and Dudel, E.G. (2007); Are *Lemna spp.* Effective Phytoremediation Agents?., Boremediation, Biodiversity Bioavailability, 1(1), 56-71.

Moronta, R., R. Mora, and E. Morales. (2006);.Response of the Microalga *Chlorella sorokiniana* to pH, Salinity and Temperature in Axenic and non Axenic Conditions. Revista De La Facultad De Agronomia, Universidad Del Zulia 23:28-43.

Muramoto, S. and Oki, Y. (1983); Removal of Some Heavy Metals From Polluted Water By Water Hyacinth (*Eichhornia crassipes*).,*Bull. Environ. Contam.Toxicol.*, 30, 170-177.

Mwegoha WJS (2008); The use of phytoremediation technology for abatement soil and groundwater pollution in Tanzania: opportunities and challenges. J. Sustain. Dev. Afr. 10(01):140-156.

Ornes, W.H., and Sutton, D.L. (1975); Removal of Phosphorus from Static Sewage Effluent By Water Hyacinth. *Hyacinth Cont. J.*, 13, 56-61.

Pilon-Smits E (2005); Phytoremediation, Annu. Rev. Plant. Biol. 56: 15- 39. Available at arjournals.annualreviews.org

Reeves, R. D., and R. R. Brooks.(1983); Hyperaccumulation of Lead and Zinc by Two Metallophytes from Mining Areas of Central Europe. Environ. Pollut. Ser. A. 31:277-285.

Reddy, K.R. (1983); Fate of nitrogen and phosphorus in a waste-water retention reservoir containing aquatic macrophytes., J. Environ. Qual., 12(1), 137-141.

Reeves R, Baker A. (2000); Metal accumulating plants. In: Raskin I, Ensley B (eds) Phytoremediation of toxic metals: using plants to clean up the environment. Wiley, New York, pp 193–229

Reddy, K.R., and Tucker, J.C. (1983); Productivity and nutrient uptake of water hyacinth, *Eichhornia crassipes* I. Effect of nitrogen source., Econ. Bot., 37, 237-247.

Reddy, K.R., Campbell, K.L., Graetz, D.A. and Portier, K.M. (1982); Use of biological filters for treating agricultural drainage effluents., J. Environ. Qual., 11, 591-595.

Reddy, K.R. and Debusk, W.F. (1985) Nutrient removal potential of selected aquatic macrophytes., J. Environ. Qual., 14, 459-462.

Rogers, H.H. and Davis, D.E. (1972); Nutrient removal by water hyacinth., Weed Sci., 20, 423- 428.

Roongtanakiat, N., Tangruangkiat, S. and Meesat, R. (2007); Utilization of vetiver grass (Vetiveria zizanioides) for removal of heavy metals from industrial wastewaters., ScienceAsia, 33, 397-403.

Salt DE, Blaylock M, Kumar NPBA, Dushenkov V, Ensley BD, Chet I, Raskin I. (1995a); Phytoremediation: a novel strategy for the removal of toxic metals from the environment using plants. Biotechnology 13: 468-475.

Saleem M.A. (2001); Industrialization and Water Pollution. Daily Dawn, Lahore. 15th April, 2001. Page Agriculture and Technology.

Schulz, C., Gelbrecht, J. and Rennert, B. (2004); Constructed Wetlands with Free Water Surface for Treatment of Aquaculture Effluents.,*J. Appl. Ichthyol.*, 20, 64-70.

Sen, A.K., Mondal, N.G. and Mondal, S. (1987); Studies of Uptake and Toxic Effects of Cr (VI) Schnoor J.L. (1997); Phytoremediation. University of Lowa, Department of Civil and Engineering, 1: 62.

Stephen, U.W., Schmidke, I, Stephan, V.W. and Scholtz, G. (1996); The Nicotianamine Molecule is Made-to-Measure for Complexation of Metal Micronutrients in Plants. Biometals 9: 84-90. on *Pistia stratiotes.*, *Water Sci. Technol.*, 7, 119-127.

Serruya, C. and Pollingher, U. (1983); *Lakes of the Warm Belt*. Cambridge University Press, Cambridge.

Sheffield, C.W. (1967); Water Hyacinth for Nutrient Removal., Hyacinth Cont. J., 27-30.

Stefani, G.D., Tocchetto, D., Salvato, M. and Borin, M. (2011); Performance of a Floating Treatment Wetland for In-Stream Water Amelioration in NE Italy., Hydrobiologia, 674, 157-167.

Singh, O.V., Labana, S., Pandey, G., Budhiraja, R., Jain, R.K. (2003); Phytoremediation: An Overview of Metallic ion Decontamination From Soil., *Appl. Microbiol. Biotechnol.*, 61, 405-412.

Suleimanov, A. Y. (1995); Conditions of Waste Fluid Accumulation at Petrochemical and Processing Enterprise Prevention of their Harm to Water Bodies", *Meditsina Truda Promyswe Nnaia Ekologila*, Vol. 12, pp. 31-36.

Trivedy, R.K. and Pattanshetty, S.M. (2002); Treatment of Dairy Waste by Using Water Hyacinth.,*Water Sci. Technol.*, 45(12), 329-334.

Truong, P. and Baker, D. (1998); Vetiver Grass System for Environmental Protection., Technical Bulletin no. 1998/1, Pacific Rim Vetiver Network, Office of the Royal Development Projects Board, Bangkok, Thailand.

Ugya, A.Y., I.M., Toma., and Abba A. (2015d) Comparative Studies on the Efficiency of Lemna minor L., Eicchornia crassipes and Pistia stratiotes in the phytoremediation of Refinery Waste Water. *Sciences World Journal* 10(3)

Ugya, A.Y. and T.S. Imam (2015). The efficiency of *Eicchornia crassipes* in the phytoremediation of waste water from Kaduna Refinery and petrochemical company. IOSR *J. Environ. Sci. Toxicol.* Food Technol., 9: 43-47.

Ugya, A.Y., (2015). The efficiency of *Lemna minor* L. in the phytoremediation of Romi stream: A case study of Kaduna refinery and petrochemical company polluted stream. J. Applied Biol. Biotechnol., 3: 11-14.

Ugya, A.Y., S.M. Tahir and T.S. Imam, (2015a). The efficiency of *Pistia stratiotes* in the phytoremediation of Romi stream: A case study of Kaduna refinery and petrochemical company polluted stream. *Int. J. Health Sci. Res.,* 5: 492-497.

Ugya, A.Y., T.S. Imam and A.S. Hassan, (2015c). The use of *Ecchornia crassipes* to remove some heavy metals from Romi stream: A case study of Kaduna Refinery and Petrochemical company polluted stream. *IOSR J. Pharm. Biol. Sci.,* 10: 43-46.

Ugya, A.Y., T.S. Imam and S.M. Tahir, (2015b). The use of *Pistia stratiotes* to remove some heavy metals from Romi stream: A case study of Kaduna Refinery and Petrochemical company polluted stream. *IOSR J. Environ. Sci. Toxicol. Food Technol.,* 9: 48-51.

Valipour, A., Raman, V.K. and Motallebi, P. (2010): Application of Shallow Pond System Using Water Hyacinth for Domestic Wastewater Treatment in the Presence of High Total Dissolved Solids (TDS) and Heavy Metal Salts., *Environ. Eng. Manage. J.,* 9(6), 853-860.

Valipour, A., Raman, V.K. and Ghole, V.S., (2011): Phytoremediation of Domestic Wastewater Using *Eichhornia crassipes., J. Environ. Sci. Eng.,* 53(2), 183-190.

Vymazal, J., Brix, H., Cooper, P.F., Haberl, R. Perfler R. and Laber, J. (1998): Removal Mechanisms and Types of Constructed Wetlands., In: Constructed Wetlands for Wastewater Treatment in Europe, Vymazal J, Brix H, Cooper PF, Green MB and Haberl, R. (eds), Backhuys Publishers, Leiden, pp. 17-66.

Von Wiren, N., Klair, S., Bansal, S., Briat, J.F., Khodr, H., Shiori, T., Leigh R.A. and, Hider, R.C. (1999): Nicotianamine chelates Both Fe(III) and Fe(II). Implications for Metal Transport in Plants.*Plant Physiol.* 119: 1107- 1114.

Vandenhove, H., Van Hees, M. and Van Winkel S. (2001): Feasibility of Phytoextraction to Clean Up Low-Level Uranium-Contaminated Soil. *Int. J. Phytoremediation* 3: 301-320.

United States Environmental Protection Agency (USEPA) (1997): Cleaning Up the Nation's Waste Sites: Markets and Technology Trends. EPA/542/R-96/005.Office of Solid Waste and Emergency Response, Washington, DC.

WHO (1993): *Guidelines for Drinking-water Quality. Volume 1 Recommendations.*2nd edition. World Health Organization, Geneva.

WHO, UNICEF (2000). Global water supply and sanitation assessment 2000 Report. USA: World Health Organization and United Nations Children's Fund.

Watanabe M (1997): Phytoremediation On the Brink of Commercialization. *Environ Sci Technol* 31:182–186

Wang, Q., Cui, Y. and Dong, Y. (2002): Phytoremediation of Polluted Waters: Potentials and Prospects of Wetland Plants., *Acta Biotechnol.*, 22,199-208.

Wooten, J.W. and Dodd, J.D. (1976): Growth of Water Hyacinths in Treated Sewage Effluent.,*Econ. Bot.*, 30, 29-37.

Xia, H. and Ma, X. (2006): Phytoremediation of ethion by water hyacinth (*Eichhornia crassipes*) from water., Bioresource Technol., 97, 1050-1054.

Zhu, Y.L., Zayed, A.M., Qian, J.H., Souza, M. and Terry, N. (1999): Phytoaccumulation of Trace Elements By Wetland Plants, II. Water Hyacinth.,*J. Environ. Qual.* 28, 339–344.

Zhao, Y.Q., Lu, J.B., Zhu, L. and Fu, Z.H. (2006): Effects of Nutrient Levels On Growth Characteristics and Competitive Ability of Water Hyacinth (*Eichhornia crassipes*), An Aquatic Invasive Plant., *Biodiversity Science*, 14, 159-164.

Zayed, A., Gowthaman, S. and Terry, N. (1998): Phytoaccumulation of Trace Elements by Wetland Plants: I. Duckweed., *J. Environ. Qual.*, 27, 715-721.

YOUR KNOWLEDGE HAS VALUE

- We will publish your bachelor's and
  master's thesis, essays and papers

- Your own eBook and book -
  sold worldwide in all relevant shops

- Earn money with each sale

Upload your text at www.GRIN.com
and publish for free